Alan McKirdy has written many popular books and book chapters on geology and related topics and has helped to promote the study of environmental geology in Scotland. His other books with Birlinn include *Set in Stone: The Geology and Landscapes of Scotland* and *Land of Mountain and Flood*, which was nominated for the Saltire Research Book of the Year prize. Before his retirement, he was Head of Knowledge and Information Management at Scottish Natural Heritage. Alan is now a freelance writer and has given many talks on Scottish geology and landscapes at book festivals and other events across the country.

The Northern Highlands

LANDSCAPES IN STONE

Alan McKirdy

BIRLINN

For Nigel and Margie Trewin

First published in Great Britain in 2019 by
Birlinn Ltd
West Newington House
10 Newington Road
Edinburgh
EH9 1QS

www.birlinn.co.uk

ISBN: 978 1 78027 608 3

British Library Cataloguing-in-Publication Data
A catalogue record for this book is available
on request from the British Library

Designed and typeset by Mark Blackadder

FRONTISPIECE.
View from Suilven summit.

Printed and bound in Britain by Latimer Trend, Plymouth

Contents

Introduction

The rocks of northern Scotland include some of the most ancient to be found anywhere on Planet Earth. They are the basement on which younger layers (strata) of rock were piled over 3,000 million years. These rocks tell of turbulent events involving collisions of continents that unleashed cataclysmic forces, creating a chain of mountains the remnants of which we see today on both sides of the Atlantic Ocean.

The far North-west Highlands were intensively studied by some of the early geological pioneers. They recognised the complexity of what they saw. In 1882, the Geological Survey of Great Britain's finest surveyors, Benjamin Peach and John Horne, were sent to prepare detailed maps and descriptions of the Assynt region. The resulting map and accompanying description, published in 1907, remains one of the classics of geological literature.

More gentle landscapes predominate in the east. Rocks from the Devonian Period (also known as the 'Old Red Sandstone') run from Duncansby Head in the north to the Black Isle and the Cromarty Peninsula in the south. These rocks carry the secrets of past life during what was to become known as the Age of Fishes. Hugh Miller, a celebrated writer, stonemason and fossil hunter from Cromarty, was the first to describe the amazing diversity of species of fossil fish in this area.

A thin sliver of younger fossil-bearing strata hugs the coast between Golspie and beyond Helmsdale. These rocks date back to Jurassic times. Shallow seas covered the area and a dense canopy of trees fringed these low-lying wetlands. As the vegetation died, decayed and became buried by subsequent layers of sand and silt, thin coal seams were formed from these organic remains.

This patchwork bedrock from different geological ages was, in more recent geological times, scoured by the passage of ice sheets during the last Ice Age that lasted for over 2 million years. This book attempts to make sense for the general reader of the episodes and events that shaped the familiar landscapes of the Northern Highlands.

The Northern Highlands through time

Period of geological time	Millions of years ago	Scotland's global position	Environments and events in Northern Scotland
Anthropocene	Last 10,000 years	57° N	This is the period when our species, *Homo sapiens,* lived and thrived in northern Scotland. Sustained populations of people altered the ecosystem by felling trees and introducing a system of farming. Thick layers of peat started to accumulate from around 6,000 years ago onwards.
Quaternary	Started 2 million years ago	Present position of 57° N	• **11,500 onwards** – the ice retreated as the climate started to warm. • **12,500 to 11,500 years ago** – the climate became very cold as the ice returned. • **14,700 to 12,500 years ago** – for a brief interlude, temperatures were similar to those of today. • **29,000 to 14,700 years ago** – all but the highest peaks were completely covered by ice. The ice sheet extended 70km westwards to the edge of the continental shelf. • **Before 29,000 years ago** and for a period approaching the last 2.7 million years, there were prolonged periods when thick sheets of ice covered the area. These advances of the ice were punctuated by warmer interludes, known as inter-glacials, when the temperatures rose to levels similar to those of today.
Neogene	2–24	55° N	Conditions were warm and temperate during these times, but temperatures fell as the Ice Age approached.
Palaeogene	24–65	50° N	Between 65 and 60 million years ago, the ancient continent of Pangaea was split asunder. Eight major volcanoes were active during these times including Skye and Mull volcanoes, but there is little evidence of these tumultuous events in northern Scotland.
Cretaceous	65–142	40° N	Sea levels rose to cover the area, but no rocks of this age are preserved here.

Period of geological time	Millions of years ago	Scotland's global position	Environments and events in Northern Scotland
Jurassic	142–205	35° N	Thick layers of sand, mud and limestone were deposited in shallow seas around the eastern margins of the Northern Highlands.
Triassic	205–248	30° N	Desert conditions prevailed across Scotland. Small deposits of Triassic strata are to be found near Golspie.
Permian	248–290	20° N	Desert conditions were widespread, but only isolated deposits of this age remain.
Carboniferous	290–354	On the Equator	'Scotland' was located at the Equator at this time, but no rocks of this age are preserved here.
Devonian	354–417	10° S	Desert conditions were widespread across the Old Red Sandstone continent. Braided streams fed the newly established Lake Orcadie that occupied much of the area during these times.
Silurian	417–443	15° S	Large upheavals created the Highlands of Scotland. Continents collided: Laurentia (including Scotland) and Avalonia (including England) became one landmass.
Ordovician	443–495	20° S	'Scotland' was located on the northern shores of the Iapetus Ocean, which had started to close by this period. The Durness limestones were laid down in this environment.
Cambrian	495–545	30° S	Sandstones and limestones were deposited along the coastline of the Laurentian continent.
Proterozoic	545–2,500	Close to South Pole	The Moine schists and Torridonian sandstones were formed during this period.
Archaean	Prior to 2,500	Possibly close to the South Pole	The Lewisian gneisses date from these times. The oldest rocks in the Lewisian complex were formed around 3,000 million years ago. The age of the Earth is around 4,540 million years.

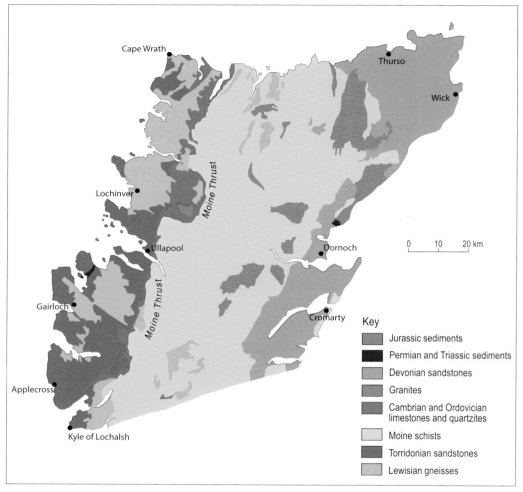

Key

▨	Jurassic sediments
■	Permian and Triassic sediments
▨	Devonian sandstones
▨	Granites
▨	Cambrian and Ordovician limestones and quartzites
▨	Moine schists
▨	Torridonian sandstones
▨	Lewisian gneisses

Geological map of the Northern Highlands. This map is one of the classics of geology. It is partly based on the work of early pioneers, including Ben Peach and John Horne from the Geological Survey, who produced a beautifully crafted map and accompanying description of the western part of the area. Lewisian gneisses, which are altered rocks from the very earliest time of Planet Earth's existence, now dated at over 3,000 million years old, are the most ancient of the rocks that make up the area. Moine schists, which occupy much of the central area of northern Scotland, originated as layers of sand and mud in a long-disappeared ocean. Dumped on top of the Lewisian gneisses are Torridonian sandstones that make up the glen of that name and also give rise to some of the most iconic mountains of the North-west Highlands, including Suilven in west Sutherland and Slioch in Wester Ross. A ribbon of limestones and sandstones runs from the north coast to Lochcarron. These strata were laid down in a coastal setting and carry the first signs of life in the area, including fossil trilobites. This relatively simple geological picture was hugely disrupted during the latter stages of a mountain-building event in which huge slices of rock were displaced westwards, as the Caledonian Mountain Belt was raised when continents collided. It's a complicated story, more of which later. Thick layers of predominately sands occupy the eastern fringe of northern Scotland. These rocks became known as the Old Red Sandstones. Younger still are the layers of Jurassic rock that fringe the coastline and hint at the presence of thicker deposits of oil-bearing strata that lie to the east under the Moray Firth.

1
Time and motion

Little of what follows will make a great deal of sense to uninitiated readers unless there is an appreciation of two main issues – geological time and the fact that the Earth's tectonic plates are in almost constant motion across the face of the planet. Grasping these concepts is essential to unravelling the geological story of the Northern Highlands.

Time

Most are aware of dim and distant events in our human history, such as the building of the twelfth-century Border abbeys or the occupation of Britain by the Romans. But, to understand our geological heritage, the clock must be wound back many millions of years. The Lewisian gneisses that underlie parts of the North-west Highlands are dated at around 3,000,000,000 years old. Such giddying numbers are difficult to grasp, but modern dating techniques have allowed scientists to use these figures with a degree of confidence. It was Professor Arthur Holmes who developed the concept of dating rocks while he was Regius Professor at Edinburgh University. His dating technique, with subsequent refinements, has helped geologists to tell the geological story of any given area from first to last, by allowing rocks and related events to be put into a reliable chronological (date) order.

Geological time is divided into manageable chunks called 'periods'. This approach helps to correlate rocks and events of similar age across the country and indeed around the world. Pages 8 and 9 of this book place the events that have left a mark on northern Scotland in date order. This sequence of events is known as a geological column. Telling the story in this 'shorthand' manner is one of the cornerstones of the subject.

Motion

It's a demonstrable fact that the earth beneath our feet is constantly on the move. The surface of the Earth is divided into seven large chunks and many smaller ones known as tectonic plates. Think of the shell of a hard-boiled egg that's been hit by a teaspoon. The resultant cracked egg-shell has a passable resemblance to the fractured nature of the Earth's outermost layer (crust).

Above. The Earth's tectonic plates move independently of each other, driven by forces deep within the planet. Friction at the margins of the plates where they jostle and rub past each other gives rise to earthquakes (shown as red dots on the world map) and occasionally associated tsunamis. The plates are very thin relative to the underlying mantle and core. If we consider the Earth as a football, the plates would be the thickness of a postage stamp stuck to the surface.

Right. The force that drives the tectonic plates across the globe is located deep within the planet. Heat radiates from the Earth's core creating convection currents in the overlying mantle. These immense forces shunt the overlying plates around the globe at an average speed of 6cm per year. So, over many millions of years, continents can move considerable distances across the face of the Earth.

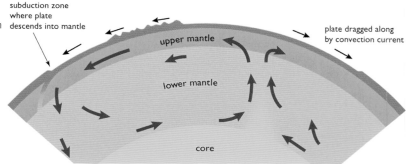

subduction zone where plate descends into mantle

upper mantle

plate dragged along by convection current

lower mantle

core

2
Early beginnings

The very oldest rocks in northern Scotland are known as the Lewisian gneisses. They are around 3,000 million (3 billion) years old and are among the most ancient rocks found anywhere in the world. This was a period of extremely turbulent geological activity. The origins of the Lewisian rocks lie deep below the surface. Localised melting took place in the Earth's mantle – the layer below the crust – and, over time, these molten rocks rose like an air bubble through syrup, ascending closer to the surface. At this time, the Earth's crust had just recently formed and complex earth movements were in progress. Also comets and asteroids had rained down on the surface of the Earth for many millions of years. But, as the Lewisian rocks comprising the early crust were forming, a greater sense of order developed. The surface of the planet thereafter divided into land and sea. The early geography of the planet bears no resemblance to that of today. The fragment of early crust that was later to form part of Scotland was made of the same

This is a re-creation of how the geography of the early world might have looked during Lewisian times. The Canadian Shield, Greenland, parts of Scotland and Scandinavia were all made from the same stuff.

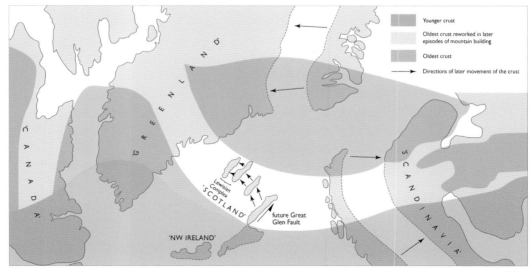

The stunning landscape of the Loch Laxford area is typical of the scenery created by Lewisian gneiss bedrock.

stuff as Canada, Greenland and northern Scandinavia. They were originally closely related in geographic position on the globe – possibly close to the South Pole.

Lewisian gneisses are a mixture of molten rocks that were originally granites and their base-rich equivalents (rocks rich in calcium, iron and magnesium) gabbros, which were both originally derived from the Earth's mantle. These hot rocks were then mixed with ancient sediments including limestones and layers of mud. Add a thick sequence of lavas that were erupted some 2,000 million years ago and that is the recipe for the rocks we see at the surface today. These Lewisian gneisses are banded, buckled and faulted, which to a trained observer is clear evidence of repeated episodes of earth movements and mountain-building in the far distant geological past. The situation was further complicated by the introduction (geologists use the word 'intrusion') of later molten sheets of dark basic rock and finally by repeated addition of pulses of molten granite.

The events that formed these rocks took place 15–30km down in the Earth's crust. So, from the time of their formation to the present day, a huge quantity of overlying rock must have been eroded away by ice, wind and water, to reveal these ancient rocks at the surface.

Ours is truly a dynamic earth! It is not surprising that these complex relationships took over a century of study to understand fully.

In terms of age, the Moine schists are the next group to be

described, which take their name from A' Mhòine in northern Sutherland. These rocks make up much of the Northern Highlands (see the geological map, page 10). Sometimes described as enigmatic, their geological history has yet to be fully unravelled, even after a century of study. These rocks were initially sands and occasional muds, laid down layer by layer in a long-disappeared ocean. Since that time, around 1,000 million years ago, the Moines have been involved in a number of mountain-building events, with each successive episode partially obliterating and 'overprinting' the characteristics of the last. The Moines contain no fossils. Granites and related rocks punched through the Moines in places, notably forming the Rogart and Helmsdale granites.

Torridonian sandstones are also commonly found along the north-west coast and immediately inland. They are of similar age to the Moines at around 1,000 million years old. These layers of sands, pebbles and cobbles were laid down by braided streams in wide river channels and ephemeral lakes. It is instructive to observe that these layers of sand were deposited directly on top of the Lewisian gneisses. This tells us that rapid erosion must have taken place to expose these ancient rocks at the surface, as Lewisian rocks are estimated to have been formed some 15–30km beneath the surface of the early crust.

The cover of Torridonian sandstone would have been continuous and thick across north-west Scotland, but erosion during the Ice Age,

This road cutting near Loch Laxford provides an excellent section through the various elements that make up the Lewisian gneiss – grey gneiss, dark pulses of magma, known as Scourie dykes, and pinkish granite.

Above. Moine rocks are often associated with bleak, featureless moorland as here at A' Mhòine in Sutherland.

Opposite top. Slioch is a mountain in two parts. The lower slopes are built from Lewisian gneiss, carved into the form of a valley with gently sloping sides. Above are Torridonian sediments deposited by rivers and streams directly onto the Lewisian surface. So we can see what the surface of the Earth looked like around a billion years ago in this part of the world! Erosion has pared back the sandstone cover to exhume this ancient surface.

Opposite bottom. Glen Torridon shows the nature of the Torridonian sediments to best effect. The layering derives from the way these rocks were formed – layer upon layer of sands and cobbles laid down by rivers and streams that flowed across the Lewisian landscape around 1,000 million years ago.

in particular, led to a considerable reduction in its extent. In places, the thickness of sandstone would have been more than 1 kilometre, so for the ice to have excavated so deeply is an impressive display of its erosive power. Many of the hills that rise from the lochan-studded platform of Lewisian gneiss are remnants of that former extensive cover of sandstone. Suilven, Stac Pollaidh and Ben More Coigach are a few examples of iconic mountains that are the isolated remnants of this once continuous blanket of sandstones.

During Cambrian times, around 455 million years earlier, some 300 million years after the deposition of the Torridonian sandstones, the land that was to become Scotland lay on the coastal plain of a large

Above. Suilven rises to a height of 731m above the surrounding landscape, creating an iconic landmark.

Right. This is how the Planet Earth looked around 500 million years ago. A wide expanse of ocean, known as the Iapetus Ocean, separated 'Scotland' from Avalonia, the land that would later become England and Wales. The continents had previously been clustered around the South Pole, but convection currents in the Earth's mantle forced the continents to split asunder. Laurentia broke free and travelled north towards the Equator. It was in this environment that the next geological episode recorded in northern Scotland took place.

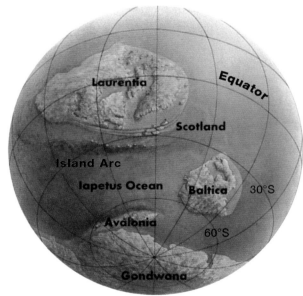

continent that included North America and Greenland. Geologists have named this landmass Laurentia.

Thick deposits of beach sands accumulated on this coastal fringe in exactly the same manner as happens today. These rocks show the

These vertical pipes are between 2cm and 5cm in length and mark the passage of burrowing creatures through the beach sands. The remains of the worms themselves have not been preserved.

The upper part of this rock face on Beinn Eighe is composed of Cambrian quartzite. This layer sits directly above Torridonian sandstone.

effects of vigorous tidal currents that swept across the tropical shallow seas and wide beaches. In places, the rocks carry the marks made by worms and other burrowing creatures that inhabited this benign environment. This is known as the 'pipe rock'.

These beach sands were later altered by heat and pressure, becoming quartzites. They are ice-white in colour and very resistant to erosion, so have become prominent features of the landscape. Many of the iconic mountains of the North-west Highlands including Arkle, Foinaven, An Teallach, Beinn Eighe, Liathach, Canisp and Quinag are capped by this distinctive rock.

A layer of silts and muds, known as the Fucoid Beds, was deposited on top of the quartzites. Some of the earliest forms of life to be

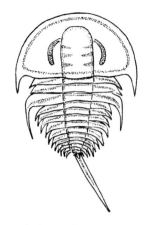

Olenellus: an early arthropod found in the Fucoid Beds, which overlie the quartzites. These strata comprise silts and muds and contain the remains of algae. They were called the Fucoid Beds as they were originally thought to contain seaweed impressions or 'fucoids'. *Olenellus* was a primitive animal, known as a trilobite, that lived on the sea floor. Similar fossils have been found in the Appalachian Mountains on the other side of the Atlantic, leading to suspicions that these areas were once much closer together. We now know that the Appalachians were part of the same mountain belt as the Scottish Highlands only to be split asunder 65 million years ago when the North Atlantic Ocean opened.

recorded anywhere in Britain are found here: more worm burrows and ancient forms of snails, cockles and sea urchins are present in these rocks. These strata were quarried during the Second World War as a source of potash. Above these thin shales are the Salterella Grits, so called because they contain remains of a fossil snail of that name.

The youngest part of the sequence of Cambrian rocks is the Durness Limestones – now formally known as the Durness Group. They take their name from a village on the north coast that has fine exposures of these rocks. The thin ribbon of limestones can be traced from this location on the north coast southwards to Lochcarron. Even when these rocks are not visible at the surface, their presence can be detected as the pastures covering these rocks take on a lush green colour.

Limestones around Elphin in south-west Sutherland turn the pastures green.

3
Continents collide

This early period of peace and tranquillity was not to last. The Iapetus Ocean, which had existed for around 250 million years, started to close due to the movement of the Earth's tectonic plates. All the layers of sediment that had accumulated on the floor of the deep ocean and in the shallower waters of the continental shelf began to be reworked to form a new mountain chain.

One of the best places to see the effects of these colossal earth movements is at Knockan Crag, north of Ullapool. This place has been a National Nature Reserve for many decades, designated for its international geological importance. An open-air visitor centre and

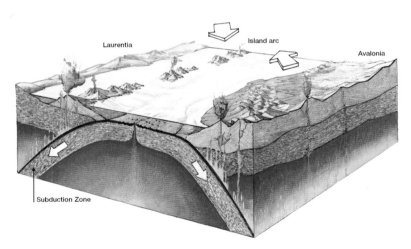

Laurentia   Island arc   Avalonia

Subduction Zone

The Iapetus Ocean began to close 480 million years ago. A line of volcanoes sitting offshore, known as an island arc, was first to collide with Laurentia as the ocean closed and a vice-like grip took hold. The layers of sand, mud, limestones and lava that had accumulated on the ocean floor were cooked, squashed, folded and faulted as unrelenting pressure of the colliding plates was applied. A towering chain of mountains was created as a result. The Caledonian Mountains stretched northwards to what became Scandinavia and southwards to form what became the Appalachians. Later tectonic readjustments of the Earth's plates split the original mountain chain apart.

Right. One of the final acts in the closure of the Iapetus Ocean was a collision between Laurentia and Baltica. Low-angled cracks, also known as thrusts (shown as a dotted line), appeared in Laurentia's leading edge, and great wedges of rock were forced westwards. Some 70km of crust was over-ridden as the pressure built. Lewisian, Torridonian, Cambrian and Moine rocks were all involved as a chaotic melange of rock wedges were stacked up along a north–south line that became known as the Moine Thrust. The north shore of Loch Glencoul is one of the best places to see evidence of this. The wedge of rock comprising Lewisian gneiss, shown above the dotted line in the image, has been driven westwards over younger strata.

Below. This is a slice through the Earth's crust that demonstrates the effects of the collision between Laurentia and Baltica. Older rocks of the Moines have been thrust over younger Cambrian strata. Along the line of the thrust, the rocks are ground up and smashed by the forces of friction. This created a puzzle that took early survey geologists many years to solve.

walking trail have been constructed to explain this global story of colliding continents (see 'Places to visit').

The closure of the Iapetus Ocean was also associated with the generation of copious amounts of granite magma. In the white heat of continent-to-continent collision, granite magmas were generated that today make up many of the landscape highs, such as Ben Loyal, Loch Ailsh and Carn Chuinneag. They ascended through the crust to a location close to the Earth's surface. They were subsequently 'unroofed' when the rock between the granite and the surface was removed by later erosion.

These granites and related rocks were injected into the early crust of the area over a long period of time. The chemical composition and component minerals of these once molten rocks also changed through time. The first bursts of igneous activity, which took place between late Ordovician and mid Silurian times, were largely concentrated on the Assynt area and further north around Ben Loyal. They are of unusual composition, notably higher in their sodium and potassium

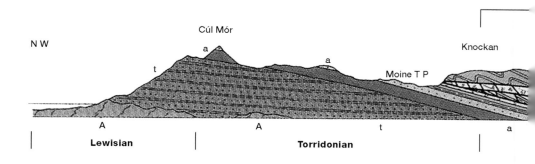

content, making them rare even in a global context. The celebrated Scottish mineralogist Matthew Heddle did much of the early study of these rocks in the 1900s.

Thin sheets of molten rock of this unusual composition were also introduced to the prominent limestone cliff near Inchnadamph in Assynt during this episode of igneous activity. These pulses of magma had the effect of altering the adjacent limestones by heat transfer, a process known as contact metamorphism (see photo on page 25).

The Helmsdale and Rogart Granites were part of a second wave of granite intrusions that spanned the time border between the Silurian

The Moine Thrust at Knockan Crag, where the older Moine rocks have been thrust over the lighter-coloured Durness Limestones.

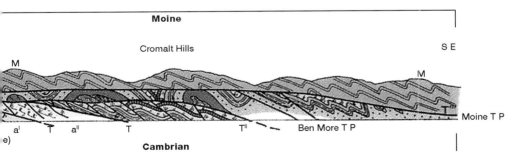

Above. The Knockan Crag visitor centre was opened in 2002 and subsequently refurbished and enhanced. Rock art, in this case a globe representing the planetary story told at Knockan, is designed to enhance the visitor experience.

Right. Ben Loyal rises to a spectacular 763m above sea level, dominating the surrounding landscape of north Sutherland.

and Devonian times. Many granites injected around the same time created what are now familiar landmarks, such as Ben Nevis and Cairngorm. The Rogart Granite is of particular interest in terms of intrusions of this age. It incorporates a few lumps of rock that are thought to come from the Earth's mantle, so this granite tapped a very deep source indeed.

Panning for gold, shed from the Helmsdale Granite, reached fever pitch in the 1840s. At its height, some 600 prospectors worked the Kildonan Burn, all looking to make their fortune. An estimated 3,500 ounces of gold have been recovered from the area since that time.

Molten sheets of magma were introduced into the Cambrian succession near Inchnadamph.

The Kildonan gold rush was in full swing in the 1840s. Early finds of gold nuggets were widely reported in the press and this swelled the number of prospectors who visited the area.

4

The Age of Fishes

After the mayhem associated with the creation of the Caledonian Mountains, a time of relative calm subsequently descended during the Devonian times around 400 million years ago. A new landmass was created by all the continental rearrangements that characterised the closure of the Iapetus Ocean. Geologists named this new, albeit now long-disappeared, land the Old Red Sandstone continent (also known as Laurussia). It was quite long lived, and the proximity of 'Scotland' to 'North America' and 'Greenland' continued until 65 million years ago, when there was another significant shift of the tectonic plates.

The Caledonian Mountains soared to heights comparable with the present-day Himalayas. The Old Red Sandstone continent was initially very unstable. Earthquakes regularly shook the land, erosion was rapid and the tops of the mountains were quickly planed down. There was no vegetation to bind the steep rock faces, so boulders

The Old Red Sandstone continent was created after the Iapetus Ocean closed. It comprised the land that would later become North America, Greenland and bits of northern Europe. 'Scotland' had moved significantly northwards but still remained south of the Equator.

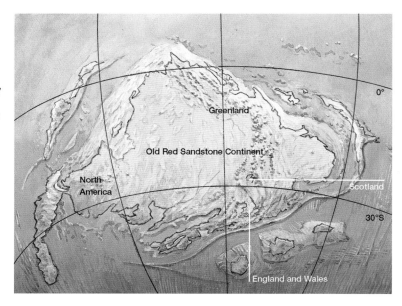

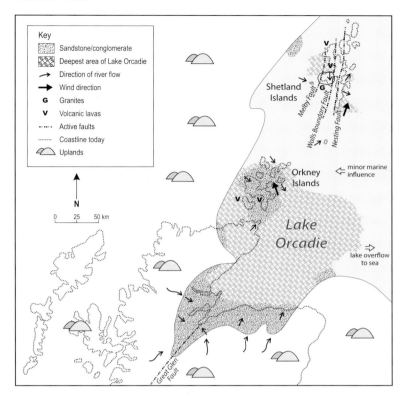

Key
- Sandstone/conglomerate
- Deepest area of Lake Orcadie
- → Direction of river flow
- ➡ Wind direction
- G Granites
- V Volcanic lavas
- ·-·-· Active faults
- ······ Coastline today
- ⌒ Uplands

N

0   25   50 km

Shetland Islands

Melby Fault

Walls Boundary Fault

G

Nesting Fault

Orkney Islands

← minor marine influence

Lake Orcadie

⇨ lake overflow to sea

Great Glen Fault

This diagram shows the full extent of Lake Orcadie, or the Orcadian Basin as it is also known. It fringed what became the Moray Firth coast and extended northwards into Orkney and Shetland. It also extended eastwards and occupied part of what is now the North Sea. The present day coastline is shown as a dotted line for reference, but the distribution of land and water was very different 400 million years ago.

tumbled down the slopes, and fast-flowing rivers transported sand, mud and pebbles to lower ground. Today, layers of boulder conglomerate provide the evidence for this interpretation. The Helmsdale Granite was exposed at the surface during this frenzied period of erosion. We know this because sandstone deposits of Devonian age are laid directly on top of the granite. This would have involved the removal of around 3km of rock cover, so the speed at which the higher ground was planed down can be reliably estimated.

The freshwater lake that became Lake Orcadie was fully established in mid-Devonian times around 400 million years ago. The extent of the lake changed from year to year depending on the amount of rainfall and other climatic factors, and occasionally the lake linked with the sea. The sediments show rapid changes from a deep-water environment to a shallower regime. At times, the lake dried out completely, so the edge of the lake changed its position from time to time. When the lake shrank, the water's edge was fringed by deserts with shifting sands blown around by ferocious winds. The lake was

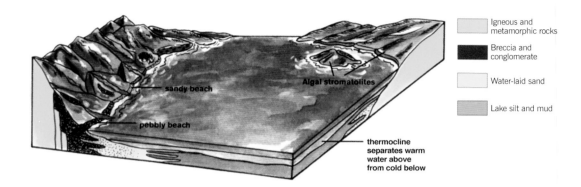

Igneous and
metamorphic rocks

Breccia and
conglomerate

Water-laid sand

Lake silt and mud

sandy beach

Algal stromatolites

pebbly beach

thermocline
separates warm
water above
from cold below

Environments and deposits in Lake Orcadie during the Middle Devonian. Lake Orcadie existed for about 10 million years. In that time, around 5km of sandstones, muds and associated sediments were laid down. Layers that indicate the lake had temporarily dried out lie close to deposits that indicate deep water, so environmental change was rapid.

never particularly deep, possibly around 80m at its deepest point. But the rate at which the lake floor subsided, perhaps driven by the weight of the accumulating layers of sand and mud, was rapid. In all, a total of 5km of sediment piled up on the lake floor.

What is especially intriguing about these rocks are the fossils they have yielded. The sediments that accumulated in Lake Orcadie preserved the fossilised remains of a wide variety of species. Famous men of science from Victorian times, notably local stonemason Hugh Miller, made their reputation from the systematic description of fossil fish from these rocks. Miller was one of the first to raise awareness of this well-preserved fossil record to a wider audience than his scientific peer group. He was, in his own words, 'an interpreter who stood between nature and the public'.

Study of these rocks has continued to the present day. Professor Nigel Trewin, late of Aberdeen University, followed in Miller's footsteps. His contribution to the study of these rocks and fossils stands comparison with Miller's earlier work. Fittingly, Nigel Trewin was President of the Friends of Hugh Miller Association for many years.

Most of the rock layers that contain the fossil remains of 400-million-year-old fish also have commercial value. Caithness flags (paving stones) have been worked for centuries and have been used as floor coverings in some of Scotland's iconic buildings, such as the Scottish Parliament. The conditions that created these flagstones were periods of gentle sediment accumulation without disturbance. The fish that died fell to the bottom of the lake and were quickly buried

Hugh Miller was a complex individual who excelled at many things. His prowess as a geologist has been justifiably celebrated. He was also a prolific writer and travelled extensively across his native Scotland. He was the first to describe a number of important fossil localities. Miller was also editor of *The Witness*, the evangelical newspaper of the Free Church of Scotland, until his tragic death at his own hand in 1856.

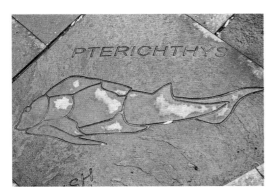

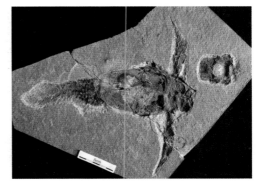

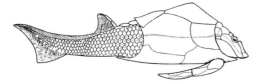

Pterichthyodes milleri is a species of fossil fish named after Hugh Miller. The image, above left, is carved into a paving slab at Miller's home in Cromarty; the image above is the fossil specimen; and the image, left, is a reconstruction of the fish as it would have been in life. It had rows of 'armour-plated' scales that protected it from the many predators that lurked in Lake Orcadie. It also had two fins that projected from its body like wings.

by sands and muds. In the lower depths of the lake, the conditions were anaerobic (free from oxygen) and, most importantly, free from scavengers. These conditions were ideal for the preservation of fish carcasses. Over time, these were fossilised in near-perfect conditions.

Professor Nigel Trewin is another titan of Old Red Sandstone geology. Working in the modern era, he published over 100 scientific papers on fossils and their associated sedimentary rocks. He had a particular fascination for the Old Red Sandstone and the fossil fishes found in these rocks. In addition to his scientific achievements, Nigel also wrote lively books on fossils that helped to bring the subject to a much wider audience. Nigel was a popular lecturer at Aberdeen University for over 40 years. His wit and boundless enthusiasm leavened even the driest of subjects.

Fish in Lake Orcadie lived in the shallows around the fringe of the lake. When they died, their bodies were transported to deeper waters where they sank to the bottom. The oxygen levels were much lower in this environment so the fish bodies remained intact until buried by sediments that rained down from above.

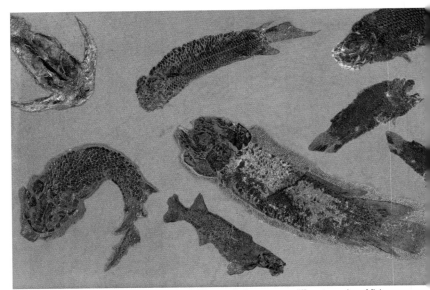

This rock slab is from Achanarras quarry in Caithness. It shows many different species of fish that died together and were fossilised as a collective. More than 15 species have been described from this one quarry. Some were predators while others fed quietly on the algae that grew in the lake.

5
A Jurassic interlude

From Devonian times, there is a gap of around 245 million years before the next significant geological episode left its mark on the area. It isn't that nothing happened here during that time, it's just that the evidence may have been sparse or has been entirely removed by later erosion. Elsewhere in Scotland, coal forests thrived during Carboniferous times, and later desert sands were widespread across the land. During this hiatus, the land that would become Scotland continued to drift from a position south of the Equator in Devonian times to 35° N at

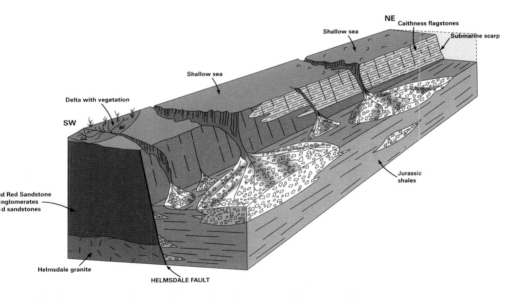

This diagram shows a slice through the land either side of the Helmsdale Fault line. The steep cliffs define the plane of the fault. There is a build-up of boulder deposits at the foot of the cliff, which also contains fossil remains of ammonites, fish and corals. At the water's edge in shallow lagoons and swamps, verdant stands of conifers and ferns grew tall. Over time, this lush vegetation cover died, decayed and became buried by a build-up of sands and muds. These layers compacted to form the coals of the Brora coalfield. Many millions of years later, these coals were worked commercially until as recently as 1975.

the onset of the Jurassic Period around 200 million years ago. These times were characterised by a dramatic global rise in sea level and many of the lower-lying landscape features of earlier times would have been covered by water. A small patch of rocks from the immediately preceding age – Triassic sands and gravels – are to be found near Golspie in eastern Sutherland.

The western extent of the Jurassic rocks is defined by the Helmsdale Fault. This sharp break in the Earth's crust runs north-east to south-west and guides the overall trend of the coastline in this part of Scotland. To the west are ancient rocks of the Moine and Helmsdale Granite. As global sea levels rose, the rocks and landscapes that had gone before simply disappeared beneath the waves.

The extent of the Jurassic strata is limited on land, but the area offshore in the Moray Firth is much larger and has been extensively drilled in pursuit of commercial oilfields. The Jurassic rocks that occur along the Brora–Helmsdale coast are also found in boreholes in wells drilled offshore.

Rock strata from the succeeding geological period – the Cretaceous – are also found extensively in offshore drill cores, but have no on-land presence, except occasional blocks dragged ashore during the Ice Age. These rocks, of a similar age to the White Cliffs of Dover, represent another ratcheting-up of sea level as Planet Earth became a greenhouse world. Temperatures were around 10°C higher than they are today. Ice from the poles completely melted, resulting in a sea level around 300m higher than at present. Much of Scotland lay under water. Cretaceous rocks were deposited across the area that we now recognise as the Moray Firth, North Sea and in the area around Shetland.

A great column of rock was dislodged as the Helmsdale Fault moved. The dislodged lump of rock, which measured around 45m in length, crashed into deep water below and was embedded in sediments of Jurassic age. When this feature, known as the 'Fallen stack of Portgower', was first described it was thought to have been a sea stack that had tipped over and landed on its side. It was only later found to be associated with the violent movements along the Helmsdale Fault.

6
The Ice Age and beyond

The high temperatures of earlier times didn't last forever. Around 7 million years ago, ice sheets re-established across Greenland and they advanced southwards to cover mid-latitude countries including Scotland. The earliest events in Scotland in this long-running age of ice are dated around 2.6 million years ago. During the Ice Age a thick layer of ice advanced and retreated in response to the changing climate.

What explains these dramatic temperature fluctuations? Our orbit around the Sun varies from circular to elliptical. This eccentric orbit has a periodicity of 100,000 years: in other words the orbit takes that amount of time to go from circular to the maximum extent of its elliptical range and back to circular again. When the Earth is furthest from the Sun, the temperature drops and ice builds up in response. But when the planet is in a completely circular orbit, and the Earth is closer to the Sun, the climate warms up and the ice cover shrinks. This warmer phase is called an inter-glacial period, and we are in one right now. They usually last for around 10,000 years and it's almost 10,000 years since the last ice sheets melted across Scotland. This means that we're nearly ready for the next advance of ice to start. But the greenhouse gases that we've added to the atmosphere since the Industrial Revolution will delay the onset of the next Ice Age by a few thousand years. The scientific consensus is that the ice will return and reclaim the whole country, but thankfully not in our lifetime.

The passage of the ice changed the landscape of northern Scotland forever. The highest mountains were planed down and great valleys were carved into the bedrock.

As the ice melted, boulders and piles of sand and gravel that had been eroded and carried along in the ice were dumped as the glaciers receded. It was a chaotic scene in the immediate aftermath of the last glaciation, but the land was soon clothed by pioneer species of plants. Within a short time, the landscape was verdant with grasses, sedges, herbs and stands of willow.

The Margerie Glacier in Alaska demonstrates the erosive power of ice. The landscapes in the Northern Highlands were created in a similar manner.

The very tops of the highest mountains of the North-west Highlands would have been visible through the thick blanket of ice and snow during the Ice Age.

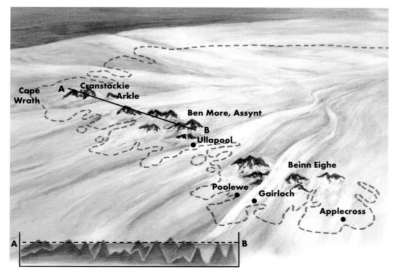

Bealach na Bà is cut into the bedrock of Torridonian sandstone and is one of the best-known landscape features of the North-west Highlands.

Boulders the size of this one and larger were transported by the ice and dumped when they could no longer be supported by the ice. These lumps of rock rejoice in the name of 'glacial erratics'. They are often different in composition from the bedrock on which they were abandoned. This indicates that they have been carried by the ice some distance and, in many cases, many kilometres from where they were carved from the bedrock.

The bone cave at Inchnadamph provides us with an exciting window into the past. Deposits that built up on the floor of the cave have preserved evidence of the ecosystem of plants and animals living in the harsh climates that existed in the immediate aftermath of the melting of the ice. Tundra conditions of freezing temperatures and biting winds prevailed at this time.

The remains of brown bears, lynx, reindeer and wolves have been excavated from the cave sediments at Inchnadamph. Recovered pollen grains provide evidence for the tree cover the existed around 8,500 years ago. Pine, hazel and birch were the dominant tree cover that surrounded the mouth of the cave.

Lynx, wolf, bear and reindeer existed in the freezing conditions after the ice melted at the end of the Ice Age.

7
Landscapes today

The Flow Country

The Flow Country covers around 400,000 hectares of Caithness and Sutherland. It consists of a thick blanket of peat that covers the bedrock to variable depths. The peat is studded by a network of pools known as *dubh lochains* (black lochs). The Flow Country was never densely forested, unlike other parts of northern Scotland, but pollen recovered from the peat confirms the presence of small stands of birch, hazel and willow here in earlier times. Accumulation of the peat started around 6,000 years ago when conditions of high rainfall and poor drainage were established. This promoted the rapid growth of Sphagnum moss – the main building block of the peatlands. Around

This aerial view across the Flow Country illustrates the nature of this peatland habitat. Black pools that can be up to 8m deep pock-mark the landscape. This largely undisturbed area is home to many rare bird species, such as waders, raptors and divers, but a limited range of plants can tolerate these extreme conditions.

Blanket bog of the
Flow Country.

4,500 years ago, there was a brief drier period when pine woodland became established on parts of the peatbog, and the remains of tree stumps can be seen to this day across the Flow Country. Peatland is recognised to be a scarce ecosystem in world terms and therefore internationally important. Strenuous efforts were made in the 1980s and 1990s to save these peatlands from significant damage caused by non-native tree planting. These efforts were largely successful. The value of this habitat as a carbon sink is also now fully recognised: carbon is trapped in the peat and unable to escape into the atmosphere as either methane or carbon dioxide, both of which are key greenhouse gases.

The machair lands

The machair lands are mainly found on the north-west coast of Scotland and are a precious resource for local crofters. The soil of these grasslands was formed from the shell fragments of sea creatures that lived in the inter-tidal zone – the area that lies between the high- and low-tide marks. They were smashed by ferocious westerly storms, and the broken shells were later washed and blown ashore. The fertile plains created by these drifting shell sands have been cultivated for

generations. The fertility of the machair is enhanced from time to time by the addition of seaweed and a rotational system of planting potatoes and hardy varieties of oats. Cattle also graze these grasslands. This managed habitat is also greatly valued as a nature conservation resource as it provides an ideal habitat for ground-nesting birds and a wealth of wild flowers.

Machair in full bloom.

The machair usually backs onto 'active' sand dunes; in other words the landscape of the dunes changes with the vagaries of the prevailing winds. Balnakeil Bay on the north coast is one of the most active sand dune systems in the area, with large steep-sided corridors cut into the dunes by relentless winds.

A dramatic sea-cliff coastline

Another natural wonder of northern Scotland is the sea cliffs which have been under attack since they were formed after the ice melted. They have been carved by the pounding surf into a variety of inter-esting coastal landforms, including stacks, arches and geos (deep clefts in the cliff-face).

The cliffs at Duncansby at the north-eastern extremity of mainland

Scotland are one of the finest examples of the destructive action of the sea. Relentless pounding by wind and waves has created a dramatic coastline cut into the sandstone bedrock. Weaknesses in the rock strata, such as faults, have been ruthlessly exploited by the sea, and the cliffs have been pared back along these planes. The sea stacks at Duncansby Head are examples of these active processes at work.

Tarbat Ness, which forms the southern headland on the Dornoch Firth, is an equally impressive cliffed coastline. The lighthouse provides additional interest.

Many dramatic coastal features have been carved out by the power of the waves and tidal currents. The sea stacks at Duncansby Head are among the most impressive.

8
Places to visit

Northern Scotland provides a cornucopia of possibilities for geological study. Nowhere else in the British Isles is geology so dominant and visible in the landscape. Every visitor has reason to stand and stare at

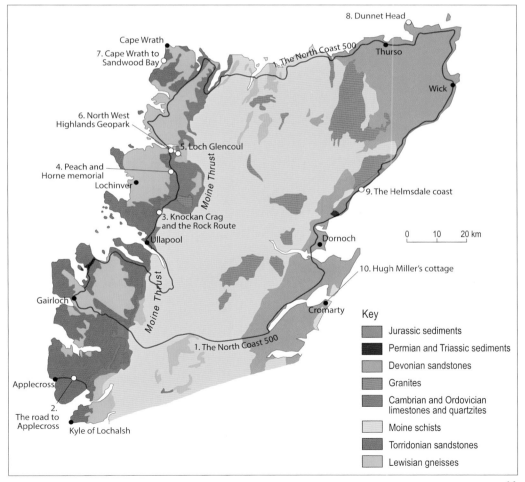

Key

- Jurassic sediments
- Permian and Triassic sediments
- Devonian sandstones
- Granites
- Cambrian and Ordovician limestones and quartzites
- Moine schists
- Torridonian sandstones
- Lewisian gneisses

8. Dunnet Head

7. Cape Wrath to Sandwood Bay

Cape Wrath

1. The North Coast 500

Thurso

Wick

6. North West Highlands Geopark

5. Loch Glencoul

4. Peach and Horne memorial

Lochinver

Moine Thrust

9. The Helmsdale coast

3. Knockan Crag and the Rock Route

Ullapool

Dornoch

10. Hugh Miller's cottage

Gairloch

Moine Thrust

Cromarty

1. The North Coast 500

Applecross

2. The road to Applecross

Kyle of Lochalsh

0   10   20 km

what nature has created here: soaring scree-covered mountains, dramatic coastlines and lochan-studded peatlands are standard fare in this part of the world. Nine OS Landranger (1:50,000 scale) maps cover the area, so identify the places you want to visit and focus map purchases on those locations. The 'Bedrock Geology UK North' map, published by the British Geological Survey, is of great assistance to anyone visiting the area.

1. The North Coast 500: this is a recently established tourist route that takes visitors around northern Scotland: from Inverness to John o' Groats and Durness to Applecross. Many of the key geological localities are accessible from this network of roads. There are dedicated publications that describe this route and the many highlights along its course.

2. The road to Applecross: this road, by way of Bealach na Bà (Pass of the Cattle), is a celebration of the geology and landscape of this

area. But don't attempt it if your mode of transport is underpowered as the climbs are steep! The road winds and wriggles its way from near Lochcarron north-westwards and, at the highest point, affords magnificent views of the landscape to the east and west. The U-shaped valley is the most obvious feature to admire. It was excavated by a valley glacier carving its way towards lower ground. The sides of the glen are made from layer upon layer of Torridonian sandstone laid down by rivers almost a billion years ago. From Applecross, proceed northwards towards Fearmore, or double back towards Lochcarron. You won't be disappointed.

3. Knockan Crag and the Rock Route: these celebrate the Assynt landscape and particularly the Moine Thrust. A circular walk takes visitors to all the key points of the site. The interpretation is aimed at those who want to know more about the local geology and appreciate some of its complexities.

4. Peach and Horne memorial: this commemorates the heroic contribution made by the early Geological Survey geologists Ben Peach and John Horne. With colleagues, they undertook a comprehensive mapping exercise of the Assynt area. This work solved many of the conundrums that this area of very complex geology previously posed. This monument is a fitting memorial to their work.

The memorial to Peach and Horne erected by the Edinburgh Geological Society.

Above. The view across the Assynt landscape from the Knockan Crag visitor centre.

Right. This is plate tectonics at a landscape scale. Great wedges of rock have been driven westwards as continents collided. The gently-dipping prominent line in the landscape is known as the Glencoul Thrust.

5. The north shore of Loch Glencoul: this is one of the best places to see the evidence for colossal earth movements associated with the Moine Thrust at a landscape scale.

6. North West Highlands Geopark: UNESCO Geopark status is given to areas of outstanding geological heritage value. Geoparks tell the geological story of their area through interpretation and educational initiatives. The North West Highlands Geopark operations are based in the old Unapool school building. During the season, the café serves hot drinks and cakes and has an associated exhibition about the rocks and landscapes of the area. The staff lead guided walks and deliver lectures to the public. It is well worth a visit.

7. Walk from Cape Wrath to Sandwood Bay: if wilderness walks are your thing, then the trek from Cape Wrath to Sandwood Bay is one of the best to be found anywhere in Britain. It doesn't get much more remote than this. Cape Wrath is the most north-westerly point on mainland Britain. The cliffs are steep, the weather will probably be extremely bracing or worse, but the landscapes and feeling of isolation are exhilarating. It's verging on an extreme adventure, so make sure that your plans are known by others before you set off. Alternatively walk to Sandwood Bay from the south, starting at Sheigra.

46

Sandwood Bay.

8. Dunnet Head: this is built from Old Red Sandstone and carved by the sea into steep cliffs and stacks. The cliffs are impressive to the east and west of the Dunnet Head lighthouse, but proceed with caution.

Dunnet Head.

9. The Helmsdale coast and southwards to Brora and Golspie: this area provides the best exposure of the Jurassic rocks in northern Scotland. The railway line hugs the coastline from Helmsdale to Brora, so safe access to the coast between those towns is limited. But, from Helmsdale, the coastal section provides access to Middle Jurassic sediments that are folded and faulted by later earth movements. Around Brora are sediments of a similar age with coal seams that were worked for hundreds of years.

10. Hugh Miller's cottage: this cottage, a 'must see' destination, is in the village of Cromarty on the Black Isle and is managed by the National Trust for Scotland. It celebrates the life, times and collected memorabilia of this great man of science. The rooms that he and his family inhabited are preserved for posterity and evoke the simplicity of life in Victorian times – sleep, work, eat and repeat. Miller leavened this routine with his fossil-collecting activities and regular church-going. His extensive collection of rocks and fossils was curated by Professor Trewin and is on display.

Hugh Miller's cottage.

Acknowledgements and picture credits

Thanks are due to Professor Stuart Monro OBE FRSE and Moira McKirdy MBE for their comments and suggestions on the various drafts of this book. I also thank Debs Warner, Mairi Sutherland, Andrew Simmons and Hugh Andrew from Birlinn Ltd for their support and direction. Mark Blackadder's book design is up to his usual high standard. Scottish Natural Heritage, in association with the British Geological Survey, published the *Landscape Fashioned by Geology* series that was the precursor to the new *Landscapes in Stone* titles. I thank them both for their permission to use some of the original artwork and photography in this book. John Mendum, Jon Merritt and Alan McKirdy wrote the original text for *The Northwest Highlands* and Clive Auton, John Merritt and Kathryn Goodenough wrote the original text for *Moray and Caithness*. Both influenced aspects of this new work. I dedicate this book to Professor Nigel Trewin (1944–2017) and his wife Margie. Nigel was an inspirational teacher during my student years at Aberdeen University in the early 1970s. He subsequently commented on, and improved, my contribution to *Land of Mountain and Flood* published by Birlinn. Margie assisted Nigel in producing the fourth edition of *Geology of Scotland*, which remains the standard reference work on Scottish geology. I was pleased when Nigel commissioned me to write a chapter on environmental geology for inclusion in that book.

Picture credits

2–3 Guillem Lopez/Alamy Stock Photo; 6 Maria Uspenskaya/Shutterstock.com; 10: © pixocreative.com; 12 (top) drawn by Robert Nelmes, (lower) drawn by Jim Lewis; 13 © pixocreative.com; 14 Lorne Gill/SNH; 15 Alan McKirdy; 16 Lorne Gill/SNH; 17 (top) Lorne Gill/SNH, (lower) Lorne Gill/SNH; 18 (top) Jaroslav Sekeres/Shutterstock.com, (lower) Richard Bonson/SNH; 19 (lower) John Gordon; 20 (top) reproduced with permission from Nigel Trewin, *Scottish Fossils*, Dunedin Academic Press, 2013, (lower) Lorne Gill/SNH; 22 (top) Lorne Gill/SNH; 23 Lorne Gill/SNH; 24 (top) I. Rottlaender/Shutterstock.com, (lower) Sara Winter/Shutterstock.com; 25 (top) Lorne Gill/SNH, (lower) © Diane Sutherland; 26 drawn by Jim Lewis; 27 © pixocreative.com; 28 drawn by Robert Nelmes; 29 reproduced with permission from Nigel Trewin, *Scottish Fossils*, Dunedin Academic Press, 2013; 30 (left) courtesy of Marge Trewin, (right) original artwork by Nigel Trewin, (lower) CP19/042 BGS © UKRI 2019; 31 Craig Ellery; 32 Arterra Picture Library/Alamy Stock Photo; 34 (top) Joshua Lombard, (lower) Craig Ellery; 35 (top) Stefano_Valeri/Shutterstock.com, (lower) Lorne Gill/SNH; 36 (top) Lorne Gill/SNH, (collage) Joe Dailly/Shutterstock.com (lynx), Bildagentur Zoonar GmbH/Shutterstock.com (wolf), zokru/Shutterstock.com (bear), MJSquared Photography/Shutterstock.com (reindeer); 37 Steve Moore/SNH; 38 Lorne Gill/SNH; 39 Natasa Kirin/Shutterstock.com; 40 Lorne Gill/SNH; 41 © pixocreative.com; 42 Matt Buckley/Shutterstock.com; 43 Lorne Gill/SNH; 44–45 courtesy of North West Highlands Geopark; 44 (lower) Lorne Gill/SNH; 45 (lower) courtesy of North West Highlands Geopark; 46 (top) Tim Askew/Shutterstock.com, (lower) Chrisnoe/Shutterstock.com; 47 Alan McKirdy.